Report to Congressional Requesters

February 2012

INFORMATION TECHNOLOGY

Departments of Defense and Energy Need to Address Potentially Duplicative Investments

GAO

Accountability * Integrity * Reliability

GAO-12-241

INFORMATION TECHNOLOGY

Departments of Defense and Energy Need to Address Potentially Duplicative Investments

Highlights of GAO-12-241, a report to congressional requesters

Why GAO Did This Study

The federal government spends billions of dollars on information technology (IT) each year, with such investments accounting for at least $79 billion in fiscal year 2011. Given the size of these investments, it is important that federal agencies avoid duplicative investments when possible to ensure the most efficient use of resources. GAO has previously reported on initiatives under way to address potentially duplicative IT investments—i.e., investments providing similar functions across the government. GAO was asked to review the extent to which potentially duplicative IT investments exist within three categories at selected agencies (the Departments of Defense (DOD), Energy (DOE), and Homeland Security (DHS)) and actions these agencies are taking to address them. To accomplish this, GAO analyzed budget data on agency IT investments, reviewed agency information related to efforts to address duplication, and interviewed agency officials.

What GAO Recommends

GAO recommends that DOD and DOE report on the progress of efforts to identify and eliminate duplication, where appropriate. GAO is also recommending that DOD, DOE, and DHS correct misclassifications of investments. DOD and DHS agreed with the recommendations. DOE generally agreed with the first recommendation, but disagreed with parts of the second recommendation regarding the number of misclassified investments. However, GAO believes the number is accurate.

View GAO-12-241. For more information, contact David A. Powner at (202) 512-9286 or pownerd@gao.gov.

What GAO Found

Although the Departments of Defense (DOD) and Energy (DOE) use various investment review processes to identify duplicative investments, GAO found that 37 of its sample of 810 investments were potentially duplicative (see table). These investments account for about $1.2 billion in total information technology (IT) spending for fiscal years 2007 through 2012. For example, GAO identified four DOD Navy personnel assignment investments—one system for officers, one for enlisted personnel, one for reservists, and a general assignment system—each of which is responsible for managing similar functions. While GAO did not identify any potentially duplicative investments at the Department of Homeland Security (DHS) within its sample, DHS officials have independently identified several duplicative investments and systems.

Potentially Duplicative Investments

Department	Purpose	Number of investments	Planned and actual expenditures ($ in millions)
DOD	Acquisition Management	4	$407
	Aviation Maintenance and Logistics	2	$85
	Civilian Personnel Management	2	$504
	Contract Management	10	$58
	Housing Management	2	$5
	Personnel Assignment Management	6	$40
	Promotion Rating	2	$3
	Workforce Management	3	$109
DOE	Back-end Infrastructure	3	$1
	Electronic Records and Document Management	3	$7
Total		**37**	**$1,219**

Source: GAO analysis of agencies' data.

DOD and DOE officials offered a variety of reasons for the potential duplication, such as decentralized governance and a lack of control over certain facilities. Further complicating agencies' ability to identify and eliminate duplicative investments is that investments are, in certain cases, misclassified by function. Until agencies correctly categorize their investments, they cannot be confident that their investments are not duplicative.

DHS has taken action to improve its processes for identifying and eliminating duplicative investments. For example, through reviewing portfolios of IT investments, DHS has identified much, and eliminated some, duplicative functionality in certain investments. Additionally, DOD and DOE have recently initiated plans to address potential duplication in many of the investments GAO identified, which include consolidating or eliminating systems. While these efforts may eventually yield results, they have not yet led to the elimination of duplication. For example, while DOD and DOE have specific plans to improve their IT investment review processes, officials did not provide examples of duplicative investments that had been consolidated or eliminated. Until DOD and DOE demonstrate progress on these efforts, the agencies will be unable to provide assurance that they are avoiding investment in unnecessary systems.

_____ United States Government Accountability Office

Contents

Figures

Abbreviations

CIO	chief information officer
DHS	Department of Homeland Security
DOD	Department of Defense
DOE	Department of Energy
FEA	Federal Enterprise Architecture
IT	information technology
NARA	National Archives and Records Administration
OMB	Office of Management and Budget

February 17, 2012

Congressional Requesters

The United States government spends billions of taxpayer dollars on information technology (IT) investments each year, with such investments accounting for at least $79 billion in fiscal year 2011.[1] Given the size of these investments, it is important that federal agencies avoid investing in duplicative investments, whenever possible, to ensure the most efficient use of resources.

Last year, we issued a comprehensive report that identified federal programs or functional areas where unnecessary duplication, overlap, or fragmentation exists; the actions needed to address such conditions; and the potential financial and other benefits of doing so.[2] More recently, we reported on the Office of Management and Budget's (OMB) and federal agencies' oversight of IT investments and the initiatives under way to address potentially duplicative IT investments.[3] Specifically, we recently reported that there are hundreds of IT investments providing similar functions across the federal government. For example, agencies reported about 1,500 investments that perform general information and technology functions, about 775 supply chain management investments, and about 620 human resource management investments.

At your request, this report provides the results of our review to identify the extent to which potentially duplicative IT investments exist within these three categories.[4] Specifically, you asked us to identify potentially duplicative IT investments at selected agencies and the actions these

---

[1]As we previously reported, this amount does not include the IT investments of 58 independent executive branch agencies, including the Central Intelligence Agency, or of the legislative or judicial branches. See GAO, *Information Technology: OMB Needs to Improve Its Guidance on IT Investments,* GAO-11-826 (Washington, D.C.: Sept. 29, 2011).

[2]GAO, *Opportunities to Reduce Potential Duplication in Government Programs, Save Tax Dollars, and Enhance Revenue,* GAO-11-318SP (Washington, D.C.: Mar. 1, 2011).

[3]GAO-11-826.

[4]For the purposes of our analysis, we considered "duplication" to occur when two or more agencies or programs are engaged in the same activities or provide the same services to the same beneficiaries.

agencies are taking to address them. We selected for review three of the largest agencies with respect to number of investments–the Departments of Defense (DOD), Energy (DOE), and Homeland Security (DHS).

To identify potentially duplicative IT investments within each of the selected agencies, we analyzed a subset of investment data from OMB's exhibit 53 to identify investments with similar functionality.[5] Specifically, we reviewed 810, or 11 percent, of the approximately 7,200 IT investments federal agencies report to OMB through the exhibit 53. Our review represents approximately 24 percent of DOD's IT portfolio in terms of the number of investments that they report to OMB, 19 percent of DOE's, and 16 percent of DHS's. We then reviewed the name and narrative description of each investment's purpose to identify similarities among related investments within each agency (we did not review investments across agencies).[6] This formed the basis of establishing groupings of similar investments. We discussed the groupings with each of the selected agencies, and we obtained further information from agency officials and reviewed and assessed agencies' rationales for having multiple systems that perform similar functions. Additionally, when analyzing each investment's description, we compared the investment's designated Federal Enterprise Architecture (FEA)[7] primary category and sub-category with OMB's definitions for each FEA primary category and sub-category and determined whether the investment was placed in the correct FEA category. We obtained additional information from agency officials about these discrepancies. We also interviewed officials to discuss actions agencies have taken to address the potentially duplicative investments and reviewed supporting documentation.

---

[5]The exhibit 53 identifies all IT projects—both major and non-major—and their associated costs within a federal organization. Information included on agency exhibit 53s is designed, in part, to help OMB better understand what agencies are spending on IT investments.

[6]Certain investments were not placed in groups because the investment descriptions were too broad. Additionally, IT investments identified as Funding Contributions were not included, since they are managed by other agencies.

[7]The FEA is intended to provide federal agencies and other decision-makers with a common frame of reference or taxonomy for informing agencies' individual enterprise architecture efforts and their planned and ongoing investment activities, and to do so in a way that identifies opportunities for avoiding duplication of effort and launching initiatives to establish and implement common, reusable, and interoperable solutions across agency boundaries.

We conducted this performance audit from June 2011 to February 2012 in accordance with generally accepted government auditing standards. Those standards require that we plan and perform the audit to obtain sufficient, appropriate evidence to provide a reasonable basis for our findings and conclusions based on our audit objectives. We believe that the evidence obtained provides a reasonable basis for our findings and conclusions based on our audit objective. See appendix I for a complete description of our objective, scope, and methodology.

Background

Information technology should enable government to better serve the American people. However, according to OMB, despite spending more than $600 billion on IT over the past decade, the federal government has achieved little of the productivity improvements that private industry has realized from IT.[8] Too often, federal IT projects run over budget, behind schedule, or fail to deliver promised functionality. In combating this problem, proper oversight is critical. Both OMB and federal agencies have key roles and responsibilities for overseeing IT investment management. OMB is responsible for working with agencies to ensure investments are appropriately planned and justified. Additionally, each year, OMB and federal agencies work together to determine how much the government plans to spend on IT projects and how these funds are to be allocated.

Required Roles and Responsibilities for IT Investment Oversight

Congress enacted several laws to assist the federal government in better managing IT investments. The three key laws are the Paperwork Reduction Act of 1995,[9] the Clinger-Cohen Act of 1996,[10] and the E-Government Act of 2002:[11]

- The Paperwork Reduction Act of 1995 specified OMB and agency responsibilities for managing information resources, including the management of IT. Among its provisions, this law established agency responsibility for assessing and managing the risks of major

---

[8]OMB, *25 Point Implementation Plan to Reform Federal Information Technology Management* (Washington, D.C.: December 2010).

[9]44 U.S.C. § 3501 et seq.

[10]40 U.S.C. § 11101 et seq.

[11]Pub. L. No. 107-347 (Dec. 17, 2002).

information systems initiatives.[12] It also required that OMB develop and oversee policies, principles, standards, and guidelines for federal agency IT functions, including periodic evaluations of major information systems.

- The Clinger-Cohen Act of 1996 placed responsibility for managing investments with the heads of agencies and established chief information officers (CIO) to advise and assist agency heads in carrying out this responsibility. Additionally, this law required OMB to establish processes to analyze, track, and evaluate the risks and results of major capital investments in information systems made by federal agencies and report to Congress on the net program performance benefits achieved as a result of these investments.

- The E-Government Act of 2002 established a federal e-government initiative, which encouraged the use of web-based Internet applications to enhance the access to and delivery of government information and service to citizens, to business partners, to employees, and among agencies at all levels of government. The act also required OMB to report annually to Congress on the status of e-government initiatives. In these reports, OMB is to describe the administration's use of e-government principles to improve government performance and the delivery of information and services to the public.

OMB's IT Oversight Mechanisms

OMB uses the following mechanisms to help it fulfill its required oversight responsibilities of federal IT spending during the annual budget formulation process.

---

[12]According to OMB guidance, a major investment is a system or acquisition requiring special management attention because of its importance to the mission or function of the agency, a component of the agency, or another organization; is for financial management and obligates more than $500,000 annually; has significant program or policy implications; has high executive visibility; has high development, operating, or maintenance costs; is funded through other than direct appropriations; or is defined as major by the agency's capital planning and investment control process.

- OMB requires 27 federal departments and agencies[13] to provide information related to their IT investments, including agency IT investment portfolios (called exhibit 53s) and capital asset plans and business cases (called exhibit 300s).[14]

- In June 2009, OMB publicly deployed the IT Dashboard, which is intended to display near real-time information on the cost, schedule, and performance of all major IT investments. For each major investment, the Dashboard provides performance ratings on cost and schedule, a CIO evaluation, and an overall rating. The CIO evaluation is based on his or her evaluation of the performance of each investment and takes into consideration multiple variables. This evaluation is to be updated when new information becomes available that would affect the assessment of a given investment. The CIO also has the ability to provide written comments regarding the status of each investment. The Dashboard replaced OMB's Management Watch List and High-Risk List, which were previously used to highlight poorly planned or poorly performing investments on a quarterly basis. As of August 2011, the Dashboard displayed information on the cost, schedule, and performance of 797 major federal IT investments at 27 federal agencies.

According to OMB, the public display of investment data on the IT Dashboard is intended to allow OMB, other oversight bodies, and the general public to hold government agencies accountable for results and progress. In addition, the Dashboard allows users to download exhibit 53 data, which provide details on the more than 7,200 federal IT investments (totaling $78.8 billion in planned spending for fiscal year 2011). Figure 1 shows the number of IT investments and planned spending by federal agency.

---

[13]The 27 agencies are the Agency for International Development; the Departments of Agriculture, Commerce, Defense, Education, Energy, Health and Human Services, Homeland Security, Housing and Urban Development, the Interior, Justice, Labor, State, Transportation, the Treasury, and Veterans Affairs; the Army Corps of Engineers; the Environmental Protection Agency; the General Services Administration; the National Aeronautics and Space Administration; the National Archives and Records Administration; the National Science Foundation; the Nuclear Regulatory Commission; the Office of Personnel Management; the Small Business Administration; the Smithsonian Institution; and the Social Security Administration.

[14]The exhibit 300s provide a business case for each major IT investment and allow OMB to monitor IT investments once they are funded. Agencies are required to provide information on each major investment's cost, schedule, and performance.

Figure 1: Breakdown of Number of Federal IT Investments for Fiscal Year 2011 (as of July 2011)

Dollars in billions

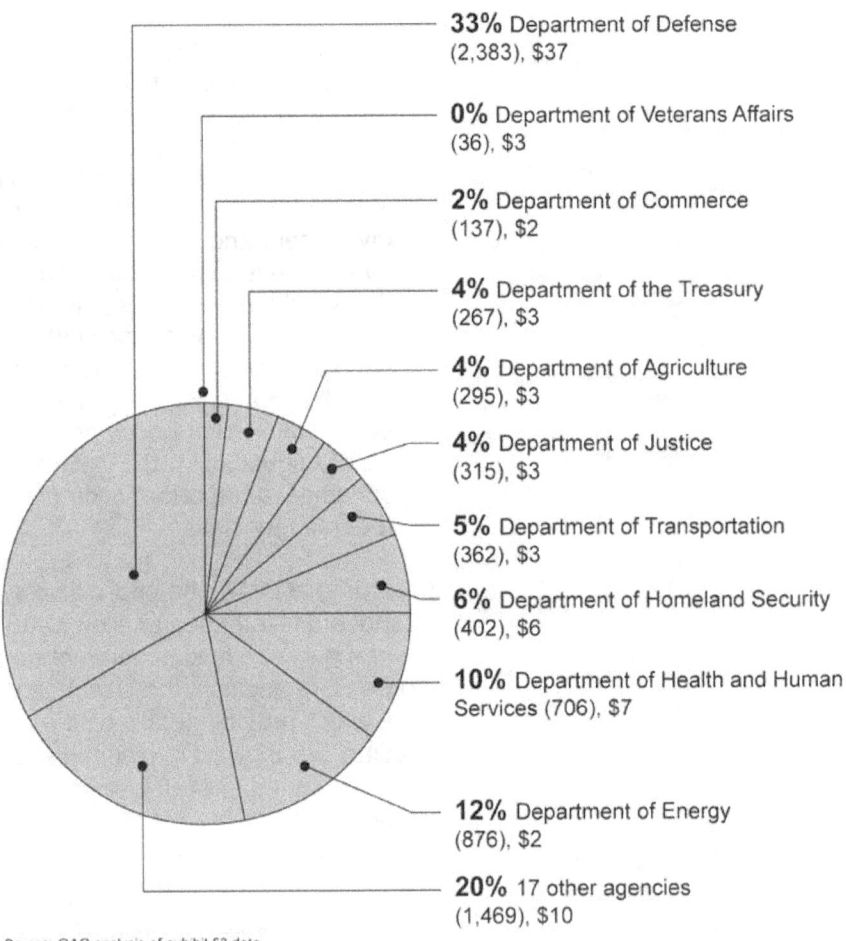

33% Department of Defense (2,383), $37

0% Department of Veterans Affairs (36), $3

2% Department of Commerce (137), $2

4% Department of the Treasury (267), $3

4% Department of Agriculture (295), $3

4% Department of Justice (315), $3

5% Department of Transportation (362), $3

6% Department of Homeland Security (402), $6

10% Department of Health and Human Services (706), $7

12% Department of Energy (876), $2

20% 17 other agencies (1,469), $10

Source: GAO analysis of exhibit 53 data.

As we have previously reported, while the IT Dashboard provides IT investment information for 27 federal agencies, it does not include any information about 61 other agencies' investments.[15] Specifically, it does not include information from 58 independent executive branch agencies

---

[15]GAO-11-826.

(such as the Securities and Exchange Commission, the Central Intelligence Agency, and the Federal Communications Commission) and 3 other agencies (such as the Legal Services Corporation). It also does not include information from the legislative or judicial branch agencies. Accordingly, we recommended that OMB specify which executive branch agencies are included when discussing the annual federal IT investment portfolio. OMB disagreed with this recommendation, stating that the agencies included in the federal IT portfolio are already identified in OMB guidance and on the IT Dashboard. However, we maintained that the recommendation had not been fully addressed because OMB officials frequently refer to the federal IT portfolio without clarifying that it does not include all agencies.

Agencies Spend Billions on Poorly Performing IT Investments

Despite required roles and responsibilities and OMB's oversight mechanisms, the federal government spends billions of dollars on poorly performing IT investments, as the following examples illustrate:

- In April 2008, due to problems identified during testing and cost overruns and schedule slippages, the Secretary of Commerce announced a redesign of the 2010 Census, resulting in a $205 million increase in life-cycle costs.

- In February 2010, the Defense Integrated Military Human Resources System was canceled after 10 years of development and approximately $850 million spent, due, in part, to a lack of strategic alignment, governance, and requirements management, as well as the overall size and scope of the effort.[16]

- In July 2010, OMB directed the National Archives and Records Administration (NARA) to halt development of its Electronic Records Archive system at the end of fiscal year 2011 (1 year earlier than planned). OMB cited concerns about the system's cost, schedule, and

---

[16]Advance Policy Questions for Testimony of Elizabeth A. McGrath to be Deputy Chief Management Officer of the Department of Defense, http://armed-services.senate.gov/statemnt/2010/03%20March/McGrath%2003-23-10.pdf (Washington, D.C.: March 2010).

performance and directed NARA to better define system functionality and improve strategic planning. Through fiscal year 2010, NARA had spent about $375 million on the system.

- In January 2011, the Secretary of Homeland Security ended the Secure Border Initiative Network program after spending about $1.5 billion because it did not meet cost-effectiveness and viability standards.[17]

- In February 2011, the Office of Personnel Management canceled its Retirement Systems Modernization program, after several years of trying to improve the implementation of this investment.[18] According to the Office of Personnel Management, it spent approximately $231 million on this investment.

- In March 2011, we reported that while DOD's Navy Next Generation Enterprise Network investment's first increment is estimated to cost $50 billion, the program was not well positioned to meet its cost and schedule estimates.[19] As such, we recommended DOD limit further investment until it conducts an interim review to reconsider the selected acquisition approach and addresses its investment management issues. DOD stated that it did not concur with the recommendation to reconsider its acquisition approach, but we maintain that without doing so, DOD cannot be sure it is pursuing the most cost-effective approach.

Additionally, as of August 2011, according to the IT Dashboard, 261 of the federal government's approximately 800 major IT investments—totaling almost $18 billion—are in need of management attention (rated

---

[17]GAO, *Border Security: Preliminary Observations on the Status of Key Southwest Border Technology Programs*, GAO-11-448T (Washington, D.C.: Mar. 15, 2011).

[18]GAO, *OPM Retirement Modernization: Longstanding Information Technology Management Weaknesses Need to Be Addressed*, GAO-12-226T (Washington, D.C.: Nov. 15, 2011).

[19]GAO, *Information Technology: Better Informed Decision Making Needed on Navy's Next Generation Enterprise Network Acquisition*, GAO-11-150 (Washington, D.C.: Mar. 11, 2011).

"yellow" to indicate the need for attention or "red" to indicate significant concerns).[20] (See fig. 2.)

Figure 2: Overall Performance Ratings of Major IT Investments on the Dashboard, as of August 2011

Dollars in billions

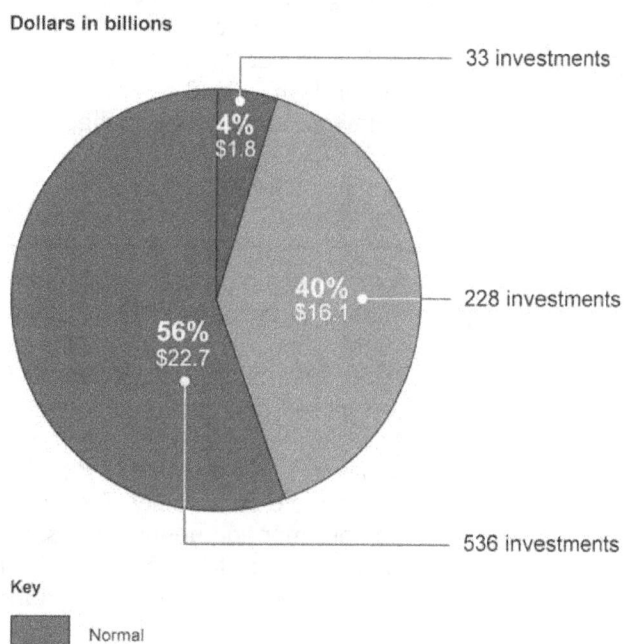

33 investments

40%
$16.1 — 228 investments

56%
$22.7

536 investments

Key

	Normal
	Needs attention
	Significant concerns

Source: OMB's IT Dashboard.

In recognizing that wasteful spending continues to plague IT investment management, OMB has recently implemented additional efforts to address this problem. These efforts include the following:

- *TechStat reviews.* In January 2010, the Federal CIO began leading reviews—known as "TechStat" sessions—of selected IT investments involving OMB and agency leadership to increase accountability and transparency and improve performance. OMB officials stated that, as

---

[20]The approximately 800 major IT investments total about $40.6 billion for fiscal year 2011.

of December 2010, 58 sessions had been held and resulted in improvements to or termination of IT investments with performance problems. For example, the June 2010 TechStat session for NARA's Electronic Records Archive investment (mentioned above) resulted in the halting of development funding pending the completion of a strategic plan. In addition, OMB has identified 26 additional high-priority IT projects and plans to develop corrective action plans with agencies at future TechStat sessions. According to the former Federal CIO, OMB's efforts to improve management and oversight of IT investments have resulted in $3 billion in savings.

- *IT reform.* In December 2010, the Federal CIO issued a *25 Point Implementation Plan to Reform Federal Information Technology Management.* This 18-month plan specified five major goals: strengthening program management, streamlining governance and improving accountability, increasing engagement with industry, aligning the acquisition and budget processes with the technology cycle, and applying "light technology" and shared solutions. As part of this plan, OMB outlined actions to, among other things, strengthen agencies' investment review boards and consolidate federal data centers. The plan stated that OMB will work with Congress to consolidate commodity IT spending (e.g., e-mail, data centers, content management systems, and web infrastructure) under agency CIOs. Further, the plan called for the role of federal agency CIOs to focus more on IT portfolio management.

Categorization of IT Investments Is Intended to Facilitate Identification of Similar IT Investments

In addition to these efforts to improve government spending on IT, avoiding unnecessary duplicative investments is critically important. In February 2002, OMB established the FEA initiative. According to OMB, the FEA is intended to facilitate governmentwide improvement through cross-agency analysis and identification of duplicative investments, gaps, and opportunities for collaboration, interoperability, and integration within and across agency programs. The FEA is composed of five "reference models" describing the federal government's (1) business (or mission) processes and functions, independent of the agencies that perform them; (2) performance goals and outcome measures; (3) means of service delivery; (4) information and data definitions; and (5) technology standards. Since the fiscal year 2004 budget cycle, OMB has required agencies to categorize their IT investments in their annual exhibit 53s according to primary function and sub-function as identified in the FEA reference models. For fiscal year 2012 submissions, agencies chose from the primary functions listed in table 1.

Table 1: FEA Primary Functions for Investments, for Fiscal Year 2012 Budget Submissions

Administrative Management	Income Security
Community and Social Services	Information and Technology Management
Controls and Oversight	Intelligence Operations
Correctional Activities	Internal Risk Management and Mitigation
Defense and National Security	International Affairs and Commerce
Disaster Management	Law Enforcement
Economic Development	Legislative Relations
Education	Litigation and Judicial Activities
Energy	Natural Resources
Environmental Management	Planning and Budgeting
Financial Management	Public Affairs
General Government	Regulatory Development
General Science and Innovation	Revenue Collection
Health	Supply Chain Management
Homeland Security	Transportation
Human Resource Management	Workforce Management

Source: OMB.

In their fiscal year 2011 submissions, agencies reported the greatest number of IT investments in Information and Technology Management (1,536 investments), followed by Supply Chain Management (777 investments), and Human Resource Management (622 investments). Similarly, planned expenditures on investments were greatest in Information and Technology Management, at about $35.5 billion. Figure 3 depicts, by primary function, the total number of investments within the 27 federal agencies that report to the IT Dashboard.

Figure 3: Number of Government IT Investments by Primary Function, as of July 2011

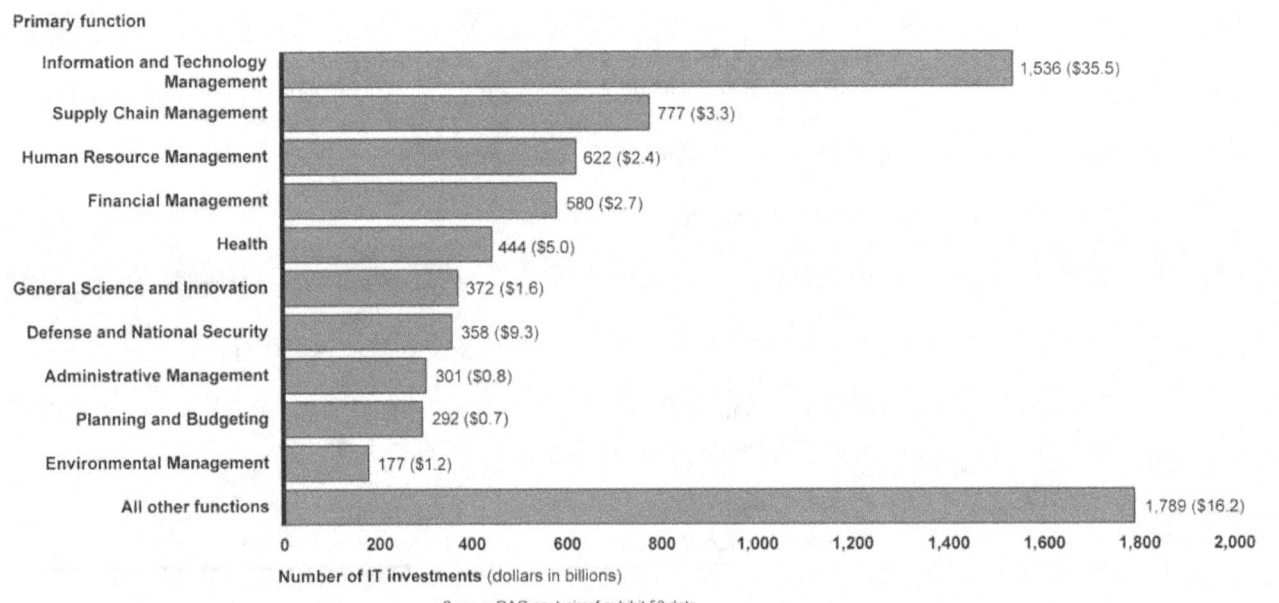

Primary function

Information and Technology Management	1,536 ($35.5)
Supply Chain Management	777 ($3.3)
Human Resource Management	622 ($2.4)
Financial Management	580 ($2.7)
Health	444 ($5.0)
General Science and Innovation	372 ($1.6)
Defense and National Security	358 ($9.3)
Administrative Management	301 ($0.8)
Planning and Budgeting	292 ($0.7)
Environmental Management	177 ($1.2)
All other functions	1,789 ($16.2)

Number of IT investments (dollars in billions)

Source: GAO analysis of exhibit 53 data.

Additionally, agencies were required to choose a sub-function for each investment related to the primary function. These sub-functions are to be selected from the business reference model. Table 2 provides examples of primary functions and their corresponding sub-functions.

Table 2: Examples of FEA Primary Functions and Corresponding Sub-Functions

Primary function	Sub-functions
Information and Technology Management	System Development
	Lifecycle/Change Management
	System Maintenance
	IT Infrastructure Maintenance
	Information Security
	Record Retention
	Information Management
	Information Sharing
	System and Network Monitoring
Supply Chain Management	Goods Acquisition
	Inventory Control
	Logistics Management
	Services Acquisition
Human Resource Management	HR Strategy
	Staff Acquisition
	Organization and Position Management
	Compensation Management
	Benefits Management
	Employee Performance Management
	Employee Relations
	Labor Relations
	Separation Management
	Human Resources Development

Source: FEA Consolidated Reference Model.

GAO Has Previously Reported on Potential Duplication and the Challenges of Identifying Duplicative Investments

During the past several years, we have issued multiple reports and testimonies and made numerous recommendations to OMB and federal agencies to identify and reduce duplication within the federal government's portfolio of IT investments.[21]

In March 2011, we reported an overview of federal programs and functional areas where unnecessary duplication, overlap, or fragmentation existed.[22] Specifically, we identified 34 areas where agencies, offices, or initiatives had similar or overlapping objectives or provided similar services to the same populations, or where government missions were fragmented across multiple agencies or programs. These areas spanned a range of government missions: agriculture, defense, economic development, energy, general government, health, homeland security, international affairs, and social services. Within and across these missions, the report touched on hundreds of federal programs, including IT programs, affecting virtually all major federal departments and agencies.

We reported that overlap and fragmentation among government programs or activities could be harbingers of unnecessary duplication. Thus, the reduction or elimination of duplication, overlap, or fragmentation could potentially save billions of tax dollars annually and help agencies provide more efficient and effective services. For example, we reported that, according to OMB, the number of federal data centers (defined as

---

[21]GAO, *IT Dashboard: Accuracy Has Improved, and Additional Efforts Are Under Way to Better Inform Decision Making*, GAO-12-210 (Washington, D.C.: Nov. 7, 2011); GAO-11-826; *Information Technology: OMB Has Made Improvements to Its Dashboard, but Further Work Is Needed by Agencies and OMB to Ensure Data Accuracy*, GAO-11-262 (Washington, D.C.: Mar. 15, 2011); *Information Technology: OMB's Dashboard Has Increased Transparency and Oversight, but Improvements Needed*, GAO-10-701 (Washington, D.C.: July 16, 2010); *Information Technology: Management and Oversight of Projects Totaling Billions of Dollars Need Attention*, GAO-09-624T (Washington, D.C.: Apr. 28, 2009); *Information Technology: OMB and Agencies Need to Improve Planning, Management, and Oversight of Projects Totaling Billions of Dollars*, GAO-08-1051T (Washington, D.C.: July 31, 2008); *Information Technology: Further Improvements Needed to Identify and Oversee Poorly Planned and Performing Projects*, GAO-07-1211T (Washington, D.C.: Sept. 20, 2007); *Information Technology: Improvements Needed to More Accurately Identify and Better Oversee Risky Projects Totaling Billions of Dollars*, GAO-06-1099T (Washington, D.C.: Sept. 7, 2006); *Information Technology: Agencies and OMB Should Strengthen Processes for Identifying and Overseeing High Risk Projects*, GAO-06-647 (Washington, D.C.: June 15, 2006).

[22]GAO-11-318SP.

data processing and storage facilities) grew from 432 in 1998 to more than 2,000 in 2010. These data centers often house similar types of equipment and provide similar processing and storage capabilities. These factors have led to concerns associated with the provision of redundant capabilities, the underutilization of resources, and the significant consumption of energy. Operating such a large number of centers places costly demands on the government. In an effort to address these inefficiencies, in February 2010, OMB launched the Federal Data Center Consolidation Initiative to guide federal agencies in consolidating data centers. Specifically, OMB and agencies plan to close over 950 of the more than 2,100 federal data centers by 2015. As of November 2011, agencies reported that a total of 149 data centers have been closed across the federal government. For example, 16 DOD data centers, 3 DOE centers, and 7 DHS centers have been closed.

In September 2011, we reported that limitations in OMB's guidance hindered efforts to identify IT duplication.[23] Specifically, OMB guidance stated that each IT investment needs to be mapped to a single functional category within the FEA to allow for the identification and analysis of potentially duplicative investments across agencies. We noted that this limits OMB's ability to identify potentially duplicative investments both within and across agencies because similar investments may be organized under different functions. Accordingly, we recommended that OMB revise guidance to federal agencies on categorizing IT investments to ensure that the categorizations are clear and that it allow agencies to choose secondary categories, where applicable, which will aid in identifying potentially duplicative investments. OMB officials generally agreed with this recommendation and stated that they plan to update the FEA reference models in the fall of 2011 to provide additional clarity on how agencies should characterize investments in order to enhance the identification of potentially duplicative investments.

We also reported that results of OMB initiatives to identify potentially duplicative investments were mixed and that several federal agencies did not routinely assess their entire IT portfolios to identify and remove or consolidate duplicative systems. Specifically, we said that most of OMB's recent initiatives have not yet demonstrated results, and several agencies did not routinely assess legacy systems to determine if they are

---

[23]GAO-11-826.

duplicative. As a result, we recommended that OMB require federal agencies to report the steps they take to ensure that their IT investments are not duplicative as part of their annual budget and IT investment submissions. OMB generally agreed with this recommendation.

Selected Agencies Have Potentially Duplicative Investments; DOD and DOE Need to Do More to Address Them

Although the Departments of Defense, Energy, and Homeland Security utilize various processes to prevent and reduce investment in duplicative programs and systems, potentially duplicative IT investments exist. Further complicating agencies' ability to identify and address duplicative investments is miscategorization of investments within agencies. Each of the agencies has recently initiated plans to address many of these investments. DHS's efforts have resulted in the identification and elimination of duplication, but DOD's and DOE's initiatives have not yet led to the elimination or consolidation of duplicative investments or functionality. Until DOD and DOE demonstrate progress on their efforts to identify and eliminate duplicative investments, and correctly categorize investments, it will remain unclear whether they are avoiding investment in unnecessary systems.

Potentially Duplicative IT Investments Exist at Selected Agencies

Each of the agencies we reviewed has IT investment management processes in place that are, in part, intended to prevent, identify, and eliminate unnecessary duplicative investments. For example, DOD's *Information Technology Portfolio Management Implementation* guide requires the evaluation of existing systems to identify duplication and determine whether to maintain, upgrade, delete, or replace identified systems. Similarly, DOE's *Guide to IT Capital Planning and Investment Control* specifies that investment business case summaries should be reviewed for redundancies and opportunities for collaboration. Additionally, according to DHS's *Capital Planning and Investment Control Guide,* proposed investments must be reviewed at the department level to determine if the proposed need is, among other things, being fulfilled by another DHS program, or already fulfilled by an existing capability.

Even with such investment review processes, of the 810 investments we reviewed,[24] we identified 37 potentially duplicative investments at DOD and DOE within three FEA categories (Human Resource Management, Information and Technology Management, and Supply Chain Management).[25] These investments account for about $1.2 billion in total IT spending for fiscal years 2007 through 2012. Specifically, we identified

- 31 potentially duplicative investments totaling approximately $1.2 billion at DOD, and

- 6 potentially duplicative investments totaling approximately $8 million at DOE.

The 37 investments comprise 12 groups of investments that appear to have duplicative purposes based on our analysis of each investment's description, budget information, and other supporting documentation from agency officials (see table 3). For example, we identified three investments at DOE that were each responsible for managing the back-end infrastructure at three different locations. We also identified four DOD Navy personnel assignment investments—one system for officers, one for enlisted personnel, one for reservists, and a general assignment system—each of which is responsible for managing similar assignment functions. Additionally, the Air Force has five investments that are each responsible for contract management, and within the Navy there are another five contract management investments. Table 3 summarizes the 12 groups of potentially duplicative investments we identified by purpose and agency. (See app. II for details on each of the 37 potentially duplicative investments.)

---

[24]We reviewed 11 percent of the total number of IT investments that agencies report to OMB through the IT Dashboard (810 of 7,227). The investments we reviewed represent approximately 24 percent of DOD's IT portfolio in terms of the number of investments reported to the Dashboard, 19 percent of DOE's, and 16 percent of DHS's. See appendix I for a complete description of our objective, scope, and methodology.

[25]Within the three selected functions, we narrowed our review to the following seven sub-functions: Benefits Management, Organization and Position Management, Employee Performance Management, Information Management, Information Security, Inventory Control, and Goods Acquisition.

Table 3: Potentially Duplicative Investments

Dollars in millions

Department	Branch or bureau	Purpose	Number of investments	Planned and actual spending fiscal years 2007-2012
DOD	Air Force	Contract Management	5	$41
	Army	Personnel Assignment Management	2	$12
	Navy	Acquisition Management	4	$407
		Aviation Maintenance and Logistics	2	$85
		Contract Management	5	$17
		Housing Management	2	$5
		Personnel Assignment Management	4	$28
		Promotion Rating	2	$3
		Workforce Management	3	$109
	DOD Enterprisewide	Civilian Personnel Management	2	$504
DOE	Energy Programs	Back-end Infrastructure	3	$1
	Energy Programs & Environmental and Other Defense Activities	Electronic Records and Document Management	3	$7
Total			**37**	**$1,219**

Source: GAO analysis of agencies' data.

We did not identify any potentially duplicative investments at DHS within our sample; however, DHS has independently identified several duplicative investments and systems. Specifically, DHS officials have identified and, more importantly, reduced duplicative functionality in four investments by consolidating or eliminating certain systems within each of these investments. DHS officials have also identified 38 additional systems that they have determined to be duplicative. For example, officials identified multiple personnel action processing systems that could be consolidated.

Officials from the three agencies reported that duplicative investments exist for a number of reasons, including decentralized governance within the departments and a lack of control over contractor facilities. For example, DOE investments for the management of back-end infrastructure are for facilities which DOE oversees but does not control. In addition, DOD officials indicated that a key reason for potential duplication at the Department of the Navy is that it had traditionally used a decentralized IT management approach, which allowed offices to develop systems independent of any other office's IT needs or acquisitions.

GAO-12-241 Information Technology Investments

Further complicating the agencies' ability to prevent investment in duplicative systems or programs is the miscategorization of investments. Among the 810 investments we reviewed, we identified 22 investments where the selected agencies assigned incorrect FEA primary functions or sub-functions.[26] Specifically, we identified 13 miscategorized investments at DOD, 4 at DOE, and 5 at DHS. Examples are as follows:

- DOD's Computer Aided Procurement System was initially categorized within the Information and Technology Management primary function, but DOD agreed that this investment should be classified within the Supply Chain Management primary function.

- DOE's Environmental Management Headquarters Central Internet Database was initially categorized within the Information and Technology Management primary function, but DOE agreed that this investment could be assigned the Environmental Management primary function and the Environmental Monitoring and Forecasting sub-function.

- DHS's Federal Emergency Management Agency—Minor Personnel/Training Systems investment was initially categorized within the Employee Performance Management sub-function, but DHS agreed that this investment should be assigned to the Human Resources Development sub-function.

Agency officials agreed that they had inadvertently miscategorized 15 of the 22 investments we identified. However, proper categorization is necessary in order to analyze and identify duplicative investments, both within and across agencies. Each improper categorization represents a possible missed opportunity to identify and eliminate an unjustified duplicative investment. Until agencies correctly categorize their investments, they cannot be confident that their investments are not duplicative and are justified, and they may continue expending valuable resources developing and maintaining unnecessarily duplicative systems.

---

[26]See appendix III for a complete listing of these investments.

Agencies Have Recently Initiated Plans to Address Potential Duplication in Many Investments, but Results Have Yet to Be Realized at DOD and DOE

DHS has taken action to improve its processes for identifying and eliminating duplicative investments, which has produced tangible results. Specifically, in 2010 and 2011, the DHS CIO conducted program and portfolio reviews of hundreds of IT investments and systems. DHS evaluated portfolios of investments within its components to avoid investing in systems that are duplicative or overlapping, and to identify and leverage investments across the department. Among other things, this effort contributed to the identification and consolidation of duplicative functionality within four investments. DHS also has plans to further consolidate systems within these investments by 2014, which is expected to produce approximately $41 million in cost savings. The portfolio reviews also contributed to the identification of 38 additional systems that are duplicative. Additionally, the DHS CIO and Chief Human Capital Officer are coordinating to streamline and consolidate the department's human resources investments. A summary of the investments for which DHS eliminated duplicative functionality and systems is provided in table 4 below.

Table 4: DHS Investments Consolidated or Eliminated to Reduce Duplicative Functionality

Investment title	Action	Cost savings estimate
DHS—Integrated Security Management System	Consolidated six personnel security-related systems into DHS's enterprisewide security suitability system.[a]	$2 million annually
Federal Emergency Management Agency—Time And Attendance Collection and Reporting	Eliminated this investment and now provides time and attendance functionality through DHS's enterprisewide time and attendance system.	$284,000 over 2 years
Homeland Security Information Network	Consolidated two DHS components' portals (e.g., the Federal Emergency Management Agency's Fire Services Portal) into the Homeland Security Information Network.[b]	$1 million over 5 years
Human Resources Information Technology	Consolidated five time and attendance systems into DHS's enterprisewide time and attendance system, as well as the Department of Agriculture's National Finance Center system.	Not available

Source: DHS.

[a]DHS reported that another personnel security-related system is scheduled to be consolidated into this investment during fiscal year 2012.

[b]DHS reported that an additional 12 portals will be consolidated into this investment before 2014. DHS officials estimate that these efforts will result in another approximately $41 million in savings.

DOD has begun taking action to address 29 of the 31 duplicative investments we identified. For example, according to DOD officials, four of the DOD Navy acquisition management investments—two for Naval Sea Systems Command and two for Space and Naval Warfare Systems

Command—will be reviewed to determine whether these multiple support systems are necessary. In addition, DOD reported that the Air Force is in the process of developing a single contract writing system to replace the five potentially duplicative investments we have identified. Moreover, the Department of the Navy has implemented an executive oversight board that is chaired by the Navy CIO, and it is now the Navy's single senior information management and technology policy and governance forum. The Department of the Navy also required all IT expenditures greater than $100,000 to be centrally reviewed and approved by the Navy CIO to ensure that they are not duplicative.[27] Officials reported that these initiatives will include the review of Navy's 22 potentially duplicative investments that we identified.

Similarly, DOE has plans under way to address each of the 6 investments we identified as potentially duplicative. Specifically, DOE officials established working groups that are addressing the two groups of duplicative investments we identified. These working groups are to address records management and back-end infrastructure, and are looking across the department to minimize redundancy in each of these areas. In addition, the CIO stated that DOE has developed a departmental strategy for electronic records management whereby a small number of approved records management applications will be identified for departmentwide use. Moreover, in a broader effort to reduce duplication across the department, in September and October 2011, DOE held technical strategic reviews, known as "TechStrat" sessions, which are aimed at exploring opportunities to consolidate DOE's commodity IT services, such as e-mail and help desk support, among the various DOE offices. The first two sessions provided opportunities for DOE bureaus to identify and share lessons learned, and established action items to improve DOE's IT investment portfolio.

While these efforts could eventually yield results, DOD's and DOE's initiatives have not yet led to the consolidation or elimination of duplication. For example, while DOD provided us with documented milestones—several of which have passed—for improving the Department of the Navy's IT investment review processes, officials did

---

[27]Under Secretary of the Navy's memo of December 3, 2010, "Department of the Navy Information Technology (IT)/Cyberspace Efficiency Initiatives and Realignment" and September 19, 2011, "Department of the Navy Secretariat Information Technology Expenditure Approval Authority (ITEAA)."

not provide us with any examples of duplicative investments that they had consolidated or eliminated. Similarly, while DOE officials have documented time frames for consolidating DOE's commodity IT services, electronic records management investments, and identity management investments, officials were unable to demonstrate that they have consolidated or eliminated unjustified duplicative investments.

Additionally, DOD does not have plans under way to address the remaining 2 of the 31 potentially duplicative investments. DOD officials stated that they do not have plans to address these investments because they do not agree that they are potentially duplicative. However, agency officials were unable to demonstrate that investing in these systems and programs was justified. Table 5 provides more information on the unaddressed potentially duplicative investments at DOD.

Table 5: Unaddressed Potentially Duplicative DOD Investments

Dollars in millions

Similar purpose	Branch	Investment title	Description	Total IT spending for fiscal years 2007-2012
Civilian Personnel Management	DOD Enterprise-wide	Executive Performance and Appraisal Tool	Civilian personnel management system that will become the enterprisewide automated solution for senior professional performance management.	$0.591
		Defense Civilian Personnel Data System	Corporate human resources system for civilian employees supporting the military departments and defense agencies.	$503.280

Source: GAO analysis of DOD data.

Table 6 summarizes the number of potentially duplicative investments for which Defense and Energy have actions under way, as well as the number of investments that remain unaddressed.

Table 6: Agency Plans to Address Potentially Duplicative Investments

Agency	Potentially duplicative investments	Plans under way to address duplication	No plans under way
DOD	31	29	2
DOE	6	6	0
Total	37	35	2

Source: GAO analysis of agencies' data.

Until DOD and DOE demonstrate, through existing transparency mechanisms such as OMB's IT Dashboard, that they are making progress in identifying and eliminating duplicative investments, it will remain unclear whether they are avoiding investment in unnecessary systems.

Conclusions

While agencies have various investment review processes in place that are partially designed to avoid investing in systems that are duplicative, we have identified 37 potentially duplicative investments at DOD and DOE. These investments account for about $1.2 billion in total IT spending for fiscal years 2007 through 2012. Given that our review covered 11 percent (810 investments) of the total number of IT investments that agencies report to OMB, it raises questions about how much more potential duplication exists.

DHS's recent efforts have resulted in the identification and consolidation of duplicative functionality in several investments and related systems. DOD and DOE have also recently initiated plans to address many investments that we identified, but these recent initiatives have not yet resulted in the consolidation or elimination of duplicative investments or functionality. Further complicating agencies' ability to prevent, identify, and eliminate duplicative investments is miscategorization of investments within agencies. Without demonstrating the progress of efforts to identity and eliminate duplicative investments, DOD and DOE will be unable to provide assurance that they are avoiding investment in unnecessary systems. Similarly, until DOD, DOE, and DHS, correctly categorize their investments, they are limiting their ability to identify opportunities to consolidate or eliminate duplicative investments.

Recommendations for Executive Action

To better ensure agencies avoid investing in duplicative investments, we recommend that the Secretary of Defense direct the CIO to take the following two actions:

- utilize existing transparency mechanisms, such as the IT Dashboard, to report on the results of the department's efforts to identify and eliminate, where appropriate, each potentially duplicative investment we have identified, as well as any other duplicative investments; and

- correct the miscategorizations for the DOD investments we identified and ensure that investments are correctly categorized in agency submissions.

We recommend that the Secretary of Energy direct the CIO to take the following two actions:

- utilize existing transparency mechanisms, such as the IT Dashboard, to report on the results of the department's efforts to identify and eliminate, where appropriate, each potentially duplicative investment we have identified, as well as any other duplicative investments; and

- correct the miscategorizations for the DOE investments we identified and ensure that investments are correctly categorized in agency submissions.

We recommend that the Secretary of Homeland Security direct the CIO to take the following action:

- correct the miscategorizations for the DHS investments we identified and ensure that investments are correctly categorized in agency submissions.

Agency Comments and Our Evaluation

We provided a draft of our report to the three departments selected for our review and to OMB. In commenting on the draft, DOD and DHS generally concurred with our recommendations. DOE generally agreed with our first recommendation and disagreed with parts of our second recommendation. In addition, OMB provided oral technical comments that we incorporated, where appropriate. Each department's comments are discussed in more detail below, and the written comments are reprinted in appendixes IV, V, and VI.

DOD's Deputy CIO for Information Management, Integration, and Technology within the Office of the Assistant Secretary of Defense for Networks and Information Integration provided written comments, which stated that the department agreed with both of our recommendations. DOD also provided technical comments, which we incorporated, where appropriate.

The Director of DHS's Departmental GAO/Office of Inspector General Liaison Office provided written comments, which stated that the department agreed with our recommendation to correct the miscategorized investments and ensure that investments are correctly categorized. Additionally, DHS provided documentation showing that the department had recently corrected the miscategorizations in response to our recommendation. The department also provided technical comments, which we incorporated as appropriate.

The DOE CIO provided written comments in which the department generally agreed with the first recommendation and disagreed with parts of the second recommendation. Regarding our first recommendation, to identify and eliminate potentially duplicative investments as appropriate, DOE generally agreed with the recommendation and stated that the Office of the CIO is committed to increasing its IT investment oversight. The department added that for the non-major investments that GAO identified as being potentially duplicative, it will update GAO on its progress through means other than the IT Dashboard, since non-major investments are not individually tracked on the Dashboard. However, DOE also indicated that it does not believe certain investments that we identified are potentially duplicative. Specifically, DOE did not agree that the two card issuance and maintenance, and three logical access control investments were potentially duplicative. Rather, it stated that the investments in these groups were listed individually on the exhibit 53 for reporting purposes, in order to show how the funding was being distributed at various locations. According to DOE, these costs were for the labor involved in deploying the technology, and could not be avoided given the separate geographical locations. We reviewed this additional information, and subsequently removed these five investments from our list of potentially duplicative investments.

Regarding our second recommendation to correct miscategorizations and ensure that investments are correctly categorized, DOE disagreed with parts of this recommendation. Specifically, DOE agreed that two of the four investments could be recategorized. However, it disagreed that the two training center investments should be recategorized, and stated that they should continue to be categorized under the Employee Performance Management FEA sub-function because of how they are funded. However, OMB guidance defines Employee Performance Management as activities that enable managers to make distinctions in performance and link individual performance to agency goals and mission accomplishment. In other words, this sub-function involves enabling managers to assess the performance of personnel—and does not involve providing training to personnel. In contrast, the Human Resources Development sub-function—which OMB guidance defines as administering, delivering, and designing employee development programs—is a more appropriate category.[28] Therefore, we maintain our position. Additionally, DOE stated that we identified only 4 miscategorized investments from its total population of 876 investments. However, this implies we reviewed all 876 investments. As stated in our report, we looked at 19 percent of DOE's reported IT investment population, or 167 investments, and identified 4 miscategorized investments from that subset.

In addition, DOE stated that in our September 2011 report we highlighted limitations in OMB's guidance regarding proper categorization of investments and further stated that, while OMB agreed to make improvements to the guidance, agencies and OMB did not have time to implement the changes before our new audit began. In our September report, we noted that, under OMB's guidance, agencies were unable to designate a secondary category, in addition to the primary category for each of the investments. However, in this report, our concern is with the accuracy of agencies' selections of the primary categories for certain investments. These are two independent concerns with investment categorization—both of which need to be addressed and are not necessarily dependent on each other. In other words, regardless of whether agencies are able to designate a secondary category, in addition to a primary category, it is still critically important that the primary category is accurate.

DOE made several additional comments that we address below:

- The department stated that it has implemented various investment review processes to help identify potentially duplicative investments and to manage these investments. We acknowledge in the report that DOE has such processes in place, and we provide examples of the department's existing IT investment management processes that are, in part, intended to prevent, identify, and eliminate duplicative investments.

---

[28]OMB, *FEA Consolidated Reference Model Document Version 2.3* (Washington, D.C., October 2007).

- DOE stated that our draft report mentions the Federal Data Center Consolidation Initiative but that we did not specifically discuss DOE's accomplishment in this area. In response, we added the number of federal data centers that DOE reportedly closed.

- The department stated that prior to the GAO audit, DOE officials realized potential duplicative investments may exist in back-end infrastructure and that a working group has been meeting regularly to identify duplicative investments and investigate the possibility of consolidating. We agree with this statement, and we acknowledge the working group's efforts in the report. However, as we report, this initiative has not yet resulted in the consolidation or elimination of duplicative investments or functionality.

- According to DOE, it had developed a departmental strategy for electronic records management whereby a small number of approved records management applications will be identified for department-wide use. It added that the three records management investments cited in our report will remain in place while the departmental strategy is being implemented. In response to this comment, we updated the report to acknowledge that the CIO stated that DOE has developed a departmental strategy, in addition to establishing an electronic records management working group. However, similar to the back-end infrastructure, these efforts have not yet resulted in the consolidation or elimination of duplicative investments or functionality, and thus, DOE may continue investing in unnecessary systems until such actions are taken.

- Lastly, DOE noted that in our report we discuss the Department's TechStrat sessions related to commodity IT services but did not discuss the TechStrat sessions conducted by its Office of Environmental Management on its major investments. We did not add this activity to the report, because supporting documentation was not provided to indicate that this session was conducted to specifically reduce duplication, rather than to review major investments with performance problems.

Finally, OMB's Chief Architect provided comments regarding the office's efforts to oversee IT investments, which we incorporated, as appropriate.

As agreed with your offices, unless you publicly announce the contents of this report earlier, we plan no further distribution until 11 days from the report date. At that time, we will send copies of this report to the appropriate congressional committees; the Secretaries of Defense, Energy, and Homeland Security; the Director of the Office of Management and Budget; and other interested parties. In addition, the report also will be available at no charge on GAO's website at http://www.gao.gov.

If you or your staff members have any questions on the matters discussed in this report, please contact me at (202) 512-9286 or pownerd@gao.gov. Contact points for our Offices of Congressional Relations and Public Affairs may be found on the last page of this report. GAO staff who made major contributions to this report are listed in appendix VII.

David A. Powner
Director, Information Technology
Management Issues

List of Requesters

The Honorable Joseph I. Lieberman
Chairman
The Honorable Susan M. Collins
Ranking Member
Committee on Homeland Security
 and Governmental Affairs
United States Senate

The Honorable Thomas R. Carper
Chairman
Subcommittee on Federal Financial Management,
 Government Information, Federal Services
 and International Security
Committee on Homeland Security
 and Governmental Affairs
United States Senate

The Honorable Darrell Issa
Chairman
The Honorable Elijah E. Cummings
Ranking Member
Committee on Oversight and Government Reform
United States House of Representatives

The Honorable Ben Quayle
United States House of Representatives

Appendix I: Objective, Scope, and Methodology

Our objective was to identify potentially duplicative information technology (IT) investments at selected agencies and actions these agencies are taking to address them. To select agencies for review, we used the Office of Management and Budget's (OMB) fiscal year 2011 exhibit 53. Specifically, we downloaded this data from OMB's IT Dashboard and used it to identify the agencies and their number of IT investments as reported on the Dashboard. We used this analysis to select for review three of the agencies with the highest number of IT investments—the Departments of Defense (DOD), Energy (DOE), and Homeland Security (DHS).

To identify potentially duplicative investments, we further narrowed our analysis of the exhibit 53 data to the largest Federal Enterprise Architecture (FEA)[1] primary functions, by number of investments. Within each of the selected primary functions, we selected the two sub-functions with the most investments. Table 7 identifies the FEA primary functions and FEA sub-functions used to select the investments for review.

Table 7: FEA Primary Functions and Sub-Functions Used to Select IT Investments

FEA primary function	FEA sub-function
Human Resource Management	Benefits Management
	Employee Performance Management[a]
	Organization and Position Management[a]
Information and Technology Management	Information Management
	Information Security
Supply Chain Management	Goods Acquisition
	Inventory Control

Source: GAO analysis of OMB data.

[a]Within the Human Resource Management function, our selection criteria resulted in a tie for the second-highest sub-function; we elected to include both of these sub functions.

This resulted in a nongeneralizable sample of 810 IT investments, which is 11 percent of the total number of IT investments that agencies report to OMB through the IT Dashboard (810 of 7,227).The investments we

---

[1]According to OMB, the FEA is intended to facilitate governmentwide improvement through cross-agency analysis and identification of duplicative investments, gaps, and opportunities for collaboration, interoperability, and integration within and across agency programs.

reviewed represent approximately 24 percent of DOD's IT portfolio in terms of number of investments that it reports to the Dashboard, 19 percent of DOE's, and 16 percent of DHS's. To determine the reliability of the data on the IT Dashboard, we reviewed recent GAO reports that identified issues with the accuracy and reliability of agency data on the IT Dashboard.[2] We determined that the data were sufficiently reliable for the purpose of this report, which was to identify selected investments to include in our review.

We then reviewed the name and narrative description of each investment's purpose to identify similarities among related investments within each agency (we did not review investments across agencies).[3] This formed the basis of establishing groupings of similar investments. We discussed the groupings with each of the selected agencies, and we obtained further information from agency officials. We also reviewed and assessed agencies' rationales for having multiple systems that perform similar functions. Additionally, when analyzing each investment's description, we compared each investment's designated FEA primary category and sub-category to OMB's definitions for each FEA primary category and sub-category and determined whether the investment was placed in the correct FEA category. We obtained additional information from agency officials about these discrepancies.

To identify the actions agencies have taken to address the potentially duplicative investments we identified, we reviewed agency documentation, such as agency memos and working group charters, and interviewed officials. We also reviewed documentation and interviewed agency officials to identify what investments were consolidated, eliminated, or modified to decrease duplication and the estimated cost savings (if available) associated with these actions.

We conducted this performance audit from June 2011 to February 2012 in accordance with generally accepted government auditing standards. Those standards required that we plan and perform the audit to obtain sufficient, appropriate evidence to provide a reasonable basis for our

---

[2]GAO-12-210, GAO-11-262, and GAO-10-701.

[3]Certain investments were not placed in groups because the investment descriptions were too broad. Additionally, IT investments identified as Funding Contributions were not included, since they are managed by other agencies.

findings and conclusions based on our audit objectives. We believe that the evidence obtained provides a reasonable basis for our findings and conclusions based on our audit objective.

Appendix II: Further Information on Potentially Duplicative Investments

The tables in this appendix provide information on the 37 investments that we identified as potentially duplicative within the three selected FEA functions (Human Resource Management, Information and Technology Management, and Supply Chain Management).[1] Specifically, we identified 31 potentially duplicative IT investments at DOD and 6 at DOE. Highlighted investments indicate the instances in which the agency does not currently have plans under way to address the potential duplication.

Table 8: Potentially Duplicative Investments at DOD

Dollars in millions

Similar purpose	Bureau—investment title	Description	FEA primary function	Total IT spending for fiscal years 2007-2012
Contract Management	Air Force–Contract Writing System	Contract writing system for weapons systems and science and technology.	Supply Chain Management	$4.663
	Air Force–Automated Contract Preparation System	Provides management and preparation of purchase requests and amendments, solicitations and amendments, offers, contracts, orders, modifications, supporting documents relating to the acquisition process, required management reports, and interfacing capabilities.	Supply Chain Management	$22.604
	Air Force–Contracting Information Database System	Online reporting tool for logistics contracting data.	Supply Chain Management	$9.952
	Air Force–Acquisition and Due In System	Single repository of information for items centrally procured at the Air Logistics Center; maintains and processes data for contracting and requirements activities from purchase requirements initiation through life cycle.	Supply Chain Management	$2.290
	Air Force–Contract Profit Reporting Systems	Provides decision support and calculation assistance and reporting functions for Air Force and Army procurement actions to DOD and other major commands and government agencies.	Supply Chain Management	$1.183

[1]Within the three selected functions, we narrowed our review to the following seven sub-functions: Benefits Management, Organization and Position Management, Employee Performance Management, Information Management, Information Security, Inventory Control, and Goods Acquisition.

GAO-12-241 Information Technology Investments

Dollars in millions

Similar purpose	Bureau—investment title	Description	FEA primary function	Total IT spending for fiscal years 2007-2012
Personnel Assignment Management	Army–Enlisted Distribution and Assignment System	Supports the management of the enlisted force to include assignments, deletions, and deferments. Users can create, validate, and modify requisitions. It provides enlisted strength management information, forecasting, and online query capability.	Human Resource Management	$11.545
	Army–Assignment Satisfaction Key	Self service web-based system that enables active Army enlisted soldiers to directly update assignment preferences and allows soldiers to volunteer for duty locations and special duty.	Human Resource Management	$0.006
Acquisition Management	Navy–Naval Sea Systems Command Acquisition Capabilities	Naval Sea Systems Command miscellaneous subsystems, projects, programs, special interest items, IT organizations, and sub-initiatives in support of acquisition capabilities not delineated elsewhere.	Supply Chain Management	$3.347
	Navy–Space and Naval Warfare Systems Command Acquisition Capabilities	Space and Naval Warfare Systems Command miscellaneous subsystems, projects, programs, special interest items, IT organizations, and sub-initiatives in support of acquisition capabilities not delineated elsewhere.	Supply Chain Management	$129.149
	Navy–Naval Sea Systems Command Systems Acquisition Management Capabilities	Naval Sea Systems Command miscellaneous subsystems, projects, programs, special interest items, IT organizations, and sub-initiatives in support of systems acquisition management capabilities not delineated elsewhere.	Supply Chain Management	$3.486
	Navy–Space and Naval Warfare Systems Command Systems Acquisition Management Capabilities	Space and Naval Warfare Systems Command miscellaneous subsystems, projects, programs, special interest items, IT organizations, and sub-initiatives in support of systems acquisition management capabilities not delineated elsewhere.	Supply Chain Management	$271.084
Aviation Maintenance and Logistics	Navy–Decision Knowledge Programming for Logistics Analysis and Technical Evaluation	Functions as an inventory management, current and historical flight, maintenance, engine, and aircraft data repository and warehouse. It is also planned to replace other logistical or tracking systems as investment funds are made available.	Supply Chain Management	$50.195
	Navy–Airborne Weapons Info System	Central repository of airborne weapons maintenance and logistics information. It also provides full life cycle management of weapons systems.	Supply Chain Management	$34.308

Dollars in millions

Similar purpose	Bureau—investment title	Description	FEA primary function	Total IT spending for fiscal years 2007-2012
Contract Management	Navy–Integrated Technical Item Management Program	System that supports the processing of procurement actions from requirements generation to the completion or termination of the contractual cycle. Includes all involved work from interactive information to generate procurement documentation.	Supply Chain Management	$10.267
	Navy–Space and Naval Warfare Systems Command Contract Information Management System	Application used to administer information related to procurement solicitations, solicitation amendments, large and small contracts, delivery orders, contract closeout actions, and simplified acquisition information.	Supply Chain Management	$0.858
	Navy–Space and Naval Warfare Systems Command Systems Center Atlantic Contract Information Management System	Internal contract management information system that provides real-time data on procurement acquisitions that are in the process of being awarded and all major activities in the contracting field supporting Space and Naval Warfare Systems Center Atlantic.	Supply Chain Management	$0.022
	Navy–Contract Data Requirements List	Part of Navy's paperless process that allows for the electronic preparation of Contract Data Requirements List required for contracting documents.	Supply Chain Management	$0.539
	Navy–Acquisition Management Automation System	Automated procurement system for the management of all procurement activities.	Supply Chain Management	$4.889
Housing Management	Navy–APPLY/SLATER	Online means for junior and senior officers to apply for housing in the Navy Reserve.	Human Resource Management	$0.671
	Navy–Commander, Navy Installations Command Manpower/Billets	Systems that support Manpower/Billet housing applications for Naval Installations Command.	Human Resource Management	$4.154
Personnel Assignment Management	Navy–Career Management System Interactive Detailing	Integrated web-based architecture framework that will allow fleet personnel to manage distributions, requisitions, and assignments.	Human Resource Management	$14.180
	Navy–Officer Assignment Information System II	Online officer personnel information and order-writing capabilities for use by officer assignment and placement personnel.	Human Resource Management	$1.014
	Navy–Enlisted Assignment information System	Online enlisted personnel information and order-writing capabilities for use by enlisted assignment and placement personnel.	Human Resource Management	$1.408
	Navy–Reserve Order Writing System	Standard Navy order-writing system for active and reserve officer and enlisted personnel.	Human Resource Management	$11.527

Dollars in millions

Similar purpose	Bureau—investment title	Description	FEA primary function	Total IT spending for fiscal years 2007-2012
Promotion Rating	Navy–Fleet Rating Identification System	Provides a comprehensive assessment of sailors and their eligibility and/or qualification for ratings or jobs for specialized skills. Additionally, it supports the management of accessions for entry-level personnel, entry-level career path, and administration of the reenlistment process.	Human Resource Management	$2.749
	Navy–Departmental Systems	Applications support the management of performance, performance evaluation, physical fitness program, human resources, personnel promotion, and the administration of recognition programs. For example, the Enlisted Selection Board System provides eligibility files for active duty and reserve senior enlisted members, and the Officer Promotion Administrative System maintains officer personnel data applicable to the promotion and selection board process.	Human Resource Management	$0.610
Workforce Management	Navy–Total Force Administration System	Family of systems that support specific functions within the hire-to-retire end-to-end business processes to include functional areas such as permanent change of station assignments, retention, mobilization, manpower planning, personnel and pay, promotion and performance, family advocacy, and civilian workforce development.	Human Resource Management	$89.601
	Navy–Manpower Models	Comprised of 13 models supporting core manpower planning processes of accessing, classifying, retaining, promoting, mobilizing, distributing, and assigning Marines.	Human Resource Management	$13.819
	Navy–Total Workforce Management System	Web-based application that is used by human resources management officials to track and manage their workforce data requirements.	Human Resource Management	$5.704
Civilian Personnel Management	DOD enterprisewide–Executive Performance and Appraisal Tool	Civilian personnel management system that will become the enterprisewide automated solution for senior professional performance management.	Human Resource Management	$0.591
	DOD enterprisewide–Defense Civilian Personnel Data System	Corporate human resources system for civilian employees supporting the military departments and defense agencies.	Human Resource Management	$503.280

Source: GAO analysis of DOD data.

Table 9: Potentially Duplicative Investments at DOE

Dollars in millions

Similar purpose	Bureau—investment title	Description	FEA primary function	Total IT spending for fiscal years 2007-2012
Back-end Infrastructure	Energy Programs–Office of Science Headquarters Back-end Infrastructure	Management of back-end infrastructure at headquarters.	Information and Technology Management	$0.250
	Energy Programs–Office of Science Oak Ridge Back-end Infrastructure	Management of back-end infrastructure at the Oak Ridge, Tennessee, field site.	Information and Technology Management	$0.648
	Energy Programs–Office of Science Chicago Back-end Infrastructure	Management of back-end infrastructure at the Chicago field site.	Information and Technology Management	$0.093
Electronic Records and Document Management	Environmental and Other Defense Activities–Environmental Management Carlsbad Field Office Electronic Records and Document Mgmt System	Electronic records and document management system that is to ensure the capture, preservation, and indexing of information created either manually or electronically in support of all Carlsbad, New Mexico, field office programs.	Information and Technology Management	$4.337
	Environmental and Other Defense Activities–Health and Safety Electronic Document Review System	Allows reviewers to track the review, redaction, and disposition of document review requests.	Information and Technology Management	$1.418
	Environmental and Other Defense Activities–Office of Legacy Management Record Management System	Includes the Office of Legacy Management Records Management System and the Hummingbird Records Management System. It also covers the operations and maintenance services for Hummingbird Records and Document Management System.	Information and Technology Management	$1.003

Source: GAO analysis of DOE data.

Appendix III: Miscategorized Investments

The tables in this appendix provide information on the 22 investments that we identified as incorrectly categorized by the selected agencies according to OMB's FEA.[1] Specifically, we identified 13 miscategorized investments at DOD (2 within Air Force, 2 within Army, 3 within Navy, and 6 enterprisewide), 4 at DOE, and 5 at DHS. Highlighted investments indicate the seven instances in which the agency did not agree that the investments were miscategorized.

Table 10: Miscategorized Air Force Investments at DOD

Investment title	Description	Original		Suggested	
		Primary function	Sub-function	Primary function	Sub-function
Agency IT Resources	Budget for IT resources that satisfy most IT hardware and software requirements, such as computers and scanners, not handled locally.	Information and Technology Management	Information Management	Information and Technology Management	IT Infrastructure Maintenance
Hill Ogden Air Logistics Center 508 Anti-Submarine Warfare Warfighting Mission Area 2	A grouping of the flight simulator training systems.	Information and Technology Management	Information Management	Human Resource Management	Human Resources Development

Source: GAO analysis of DOD data.

Table 11: Miscategorized Army Investments at DOD

Investment title	Description	Original		Suggested	
		Primary function	Sub-function	Primary function	Sub-function
Army Wide Information System Service Support	Provides resources that improve and assure the reliability of electric power and other utilities. It also supports enterprise software licensing agreements.	Information and Technology Management	Information Management	Information and Technology Management	IT Infrastructure Maintenance
Personnel Enterprise Support-Automation	IT infrastructure maintenance in support of a range of human resource activities.	Information and Technology Management	Information Management	Information and Technology Management	IT Infrastructure Maintenance

Source: GAO analysis of DOD data.

[1]Additional details of OMB's FEA can be found at this address: www.whitehouse.gov/omb/e-gov/fea.

Table 12: Miscategorized Navy Investments at DOD

Investment title	Description	Original		Suggested	
		Primary function	**Sub-function**	**Primary function**	**Sub-function**
ADT Picture Picture	An access control and security management system that offers a high performance database, detailed history, and active reporting generation. It also keeps access control records for buildings, rooms, facilities, turn-styles, doors, lockers, and equipment.	Information and Technology Management	Information Management	Information and Technology Management	Information Security
Judge Advocate General's Services System	Overall architecture for all Judge Advocate General system support services and applications.	Information and Technology Management	Information Management	Information and Technology Management	IT Infrastructure Maintenance
Secured Enterprise Access Tool	Provides single sign-on through common access card-based public key infrastructure certificates to a number of Pacific Fleet-Area of Responsibility web-based applications.	Information and Technology Management	Information Management	Information and Technology Management	Information Security

Source: GAO analysis of DOD data.

Table 13: Miscategorized Enterprisewide Investments at DOD

Investment title	Description	Original		Suggested	
		Primary function	Sub-function	Primary function	Sub-function
Computer Aided Procurement System	Application that converts telecommunications service requests and orders into telecommunications requirements that are used for vendor solicitation.	Information and Technology Management	Information Management	Supply Chain Management	Services Acquisition
Global Surface Distribution Management	Port opening capability that provides the facility, automated tools, and communication infrastructure.	Information and Technology Management	Information Management	Information and Technology Management	IT Infrastructure Maintenance
Industrial Security Facility Database	A centralized web-based platform that manages the industrial security facility clearance process.	Information and Technology Management	Information Management	Administrative Management	Security Management
Infostructure	Centrally procures IT hardware and logically consolidates certain transportation command systems. Additionally, it develops IT solutions to rapidly meet gaps in distribution processes.	Information and Technology Management	Information Management	Information and Technology Management	IT Infrastructure Maintenance
National Defense University's IT Sustainment	Funds for the day-to-day operations and maintenance of the National Defense University network, related software and its maintenance, information security and assurance of the network, and development of systems.	Information and Technology Management	Information Management	Information and Technology Management	IT Infrastructure Maintenance
Rates and Tariffs File System	Used to update telecommunications contracts information with defined tariffs and tariff charges.	Information and Technology Management	Information Management	Supply Chain Management	Services Acquisition

Source: GAO analysis of DOD data.

Table 14: Miscategorized Investments at DOE

Investment title	Description	Original		Suggested	
		Primary function	Sub-function	Primary function	Sub-function
Environmental Management Headquarters Central Internet Database	Designed to give the general public access to information about DOE's nuclear waste management and cleanup program.	Information and Technology Management	Information Management	Environmental Management	Environmental Monitoring and Forecasting
National Nuclear Security Administration Los Alamos National Laboratory Software Applications Training Center	This lab's software applications training center.	Human Resource Management	Employee Performance Management	Human Resource Management	Human Resources Development
National Nuclear Security Administration Los Alamos National Laboratory Virtual Training Center	This lab's virtual training center.	Human Resource Management	Employee Performance Management	Human Resource Management	Human Resources Development
Office of Nuclear Energy Idaho National Laboratory Classified Cyber Life Cycle Management	Provides for the management of data calls from DOE, the Inspector General, and other federal entities.	Information and Technology Management	Information Security	Information and Technology Management	Information Management

Source: GAO analysis of DOE data.

Table 15: Miscategorized Investments at DHS

Investment title	Description	Original Primary function	Original Sub-function	Suggested Primary function	Suggested Sub-function
United States Customs and Border Protection–Television	Satellite television broadcasting system that supports mission-critical programs for the Office of Training and Development.	Human Resource Management	Employee Performance Management	Human Resource Management	Human Resources Development
Federal Emergency Management Agency–Minor Personnel/Training Systems	Minor personnel and training systems such as Employee Knowledge Center and Complaints.	Human Resource Management	Employee Performance Management	Human Resource Management	Human Resources Development
United States Immigration and Customs Enforcement–Password Issuance and Control System	Supports the centralized issuance of user identification numbers and passwords to valid users of United States Immigration and Customs Enforcement application systems.	Information and Technology Management	Information Management	Information and Technology Management	Information Security
Transportation Security Administration–Online Learning Center	Provides delivery and maintenance of training records for Transportation Security Administration employees and contractors.	Human Resource Management	Employee Performance Management	Human Resource Management	Human Resources Development
United States Coast Guard–Ship Control and Navigation Training System	A ship-handling simulator used to train personnel on navigation, bridge team coordination, restricted water transits, and emergency procedures.	Human Resource Management	Employee Performance Management	Human Resource Management	Human Resources Development

Source: GAO analysis of DHS data.

Appendix IV: Comments from the Department of Defense

OFFICE OF THE ASSISTANT SECRETARY OF DEFENSE
6000 DEFENSE PENTAGON
WASHINGTON, D.C. 20301-6000

NETWORKS AND INFORMATION
INTEGRATION

JAN 2 0 2012

Mr. David A. Powner,
Director, Information Technology Management
U.S. Government Accountability Office
441 G Street, NW, Washington, DC 20548

Dear Mr. Powner,

This is the Department of Defense (DoD) response to the GAO draft report, GAO-12-241, 'INFORMATION TECHNOLOGY: Departments of Defense and Energy Need to Address Potentially Duplicative Investments,' dated December 15, 2011 (GAO Code 31 1251).

Our comments on the draft report are attached. My point of contact is Mr. Kevin Garrison, 571-372-4473, Kevin.garrison@osd.mil.

Sincerely,

David L. DeVries
Deputy Chief Information Officer for
Information Management, Integration & Technology

GAO Draft Report Dated December 15, 2011
GAO-12-241 (GAO CODE 311251)

"INFORMATION TECHNOLOGY: DEPARTMENTS OF DEFENSE
AND ENERGY NEED TO ADDRESS POTENTIALLY
DUPLICATIVE INVESTMENTS"

DEPARTMENT OF DEFENSE COMMENTS
TO THE GAO RECOMMENDATIONS

RECOMMENDATION 1: The GAO recommends that the Secretary of Defense direct the
Chief Information Officer to utilize existing transparency mechanisms, such as the information
technology dashboard, to report on the results of the department's efforts to identify and
eliminate, where appropriate, each potentially duplicative investment GAO has identified, as
well as any other duplicative investments. (See page 24/GAO Draft Report.)

DoD RESPONSE: Concur. Addition information concerning some of the referenced
investments is attached.

RECOMMENDATION 2: The GAO recommends that the Secretary of Defense direct the
Chief Information Officer to correct the miscategorizations for the DoD investments that GAO
has identified and ensure that investments are correctly categorized in agency submissions. (See
page 24/GAO Draft Report.)

DoD RESPONSE: Concur. Additional information concerning the categorization of some of
the referenced investments is attached.

Appendix V: Comments from the Department of Energy

Department of Energy
Washington, DC 20585

January 17, 2012

Mr. David A. Powner
Director, Information Technology Management Issues
U.S. Government Accountability Office

Dear Mr. Powner:

The Department of Energy (DOE) Office of the Chief Information Officer (OCIO) appreciates the opportunity to provide comments to the General Accountability Office's (GAO) Draft Information Technology (IT) Report, *Departments of Defense and Energy Need to Address Potentially Duplicative Investments*. We understand this audit was conducted to determine whether the Department invested in duplicative IT investments and are committed to addressing the actions outlined in the report.

DOE has implemented various investment review processes already to help identify potentially duplicative investments and to manage the cost, schedule, and performance of these investments. These reviews include our annual IT portfolio reviews, Quarterly Control Reviews, monthly IT Dashboard reporting, TechStat Reviews, and the Federal Data Center Consolidation Initiative (FDCCI). Through these reviews, the Department has identified opportunities for cost savings and efficiencies; however, we realize additional opportunities exist to reduce duplication and we are taking steps to eliminate these redundancies in various areas.

Management Response to Audit Findings/Comments

The draft report mentions the FDCCI; however, it does not specifically discuss DOE's accomplishment in this area. To date, we have closed three federal data centers, resulting in millions of dollars of savings, with future plans to close additional data centers. We are also exploring the use of alternative financing opportunities, such as Energy Savings Performance Contracts (ESPC), to supplement traditional appropriated funding to further consolidate and optimize data centers. We are considering an ESPC to facilitate the consolidation of data centers and server rooms at a few of our locations, as well as the transformation of our IT infrastructure to a more energy efficient and responsive computing environment, while reducing capital costs. If successful, we hope other federal agencies can benefit from this same model when optimizing their data centers.

The draft report also identified three specific categories of potentially duplicative IT investments. Additional information for each of the categories is provided below for your consideration.

1. Back-End Infrastructure Investments: Prior to the GAO audit, the Office of Science (SC) realized potential duplicative investments may exist in the Office's back-end infrastructure. The Science Information Technology Initiatives Working Group (SC ITWG) has been meeting regularly to identify duplicative investments and investigate the possibility of consolidating as much as possible from the three back-end

Printed with soy ink on recycled paper

GAO-12-241 Information Technology Investments

infrastructures. It is expected that some of the back-end infrastructure will need to remain with these locations, given that Headquarters, Chicago, and Oak Ridge are located at three geographically different locations.

2. Card Issuance and Maintenance and Logical Access Control (LAC) Investments: The necessary equipment for these investments was provided by DOE's Office of Health Safety and Security and is not duplicative since every employee and contractor at DOE SC located in Chicago, Oak Ridge, and Headquarters are required to have a badge. The costs shown in the report indicate labor costs for the deployment of this technology and this cost is not duplicative and could not be avoided given the three separate geographical locations. The three investments were listed individually on the Exhibit 53 not to show them as separate and distinct investments, but as budget line items to show where the funding was being distributed to implement cards and LAC requirements at the various locations.

3. Electronic Records and Document Management: DOE has developed a departmental strategy for electronic records management whereby a small number of approved Records Management Applications (RMAs) will be identified for Department-wide use. Due to the diverse nature of our Program missions and their associated records requirements, it is not feasible to limit the department to a single RMA solution. The three RM investments cited in GAO's report are current Program Office RM systems that will remain in place while the departmental strategy is being implemented.

On page 22 of the draft report, GAO also discusses the Department's TechStat sessions related to commodity IT services. However, the report does not discuss the TechStat sessions conducted by the Office of Environmental Management (EM) on its major investments in collaboration with the OCIO. EM is the first DOE program office to embrace the TechStat reviews as a management tool at the program office level.

Management Response to Recommendations

Recommendation 1: Utilize existing transparency mechanisms, such as the IT Dashboard, to report on the results of the department's efforts to identify and eliminate, where appropriate, each potentially duplicative investment we have identified, as well as any other duplicative investments

The OCIO will continue to actively monitor the Department's major IT investments on the IT Dashboard on a monthly basis, and utilize this tool, in addition to other internal investment review processes to identify, and if necessary, eliminate duplicative IT investments. All of the potentially duplicative investments identified by GAO in this report are non-major investments. Non-major investments are not tracked individually on the IT Dashboard, so we are unable to report our progress on addressing the potentially duplicative investments identified by GAO on the IT Dashboard. We will gladly continue to update GAO on our progress through other communication channels.

In addition, to show the Department's commitment to increasing IT investment oversight, the National Nuclear Security Administration (NNSA) recently developed the initial version of the

Executive and Governance Model for Information Technology (EGMIT) in November 2011. The EGMIT will provide a mechanism for gaining insight, exercising oversight, and influencing management of IT assets. NNSA has also started a Capital Planning and Investment Control (CPIC) awareness initiative in December 2011 to improve the quality assurance of major and non-major IT investment reporting. We are in the process of capturing all of our IT investments in an Enterprise Architecture repository in order to conduct an architectural review to ensure that future investments do not duplicate, interfere, or contradict existing IT investments.

**Recommendation 2: Correct the miscategorizations for the DOE investments we identified
and ensure that investments are correctly categorized in agency submissions**

On page 16 of the draft report, GAO references their September 2011 report which was not issued prior to this current audit commencing. The report highlighted the limitations in the Office of Management and Budget's (OMB) guidance that have hindered efforts to identify IT duplication. Current OMB requirements to map to a single functional category within the FEA make it difficult to identify potentially duplicative investments. OMB agreed to revise their guidance on characterizing IT investments to ensure categorizations are clear and that it would allow agencies to choose second categories. OMB agreed to make these guidance changes in the fall of 2011 to provide clarity to agencies in characterizing investments. Before they could do this and agencies could react, the current audit was started.

The OCIO continues to ensure investments are categorized correctly in agency submissions through the Department's CPIC and NNSA's EGMIT processes. Prior to submitting the Exhibit 53A to the OMB each year, the OCIO works closely with chief enterprise architects from each of the Program and Staff Offices to review the reported Federal Enterprise Architecture Primary Functions, Sub-functions, and Segment codes assigned to each investment. For this audit, the OCIO reviewed approximately 876 investments to ensure appropriate mapping of investments in the Budget Year (BY) 2012 Exhibit 53 submission. Only four of the 876 investments were identified as miscategorized, and we do not agree with all of GAO's recommended reclassifications.

Regarding the two NNSA investments highlighted in the GAO report, 1) NNSA Los Alamos National Laboratory Software Training Center and 2) NNSA Los Alamos National Laboratory Virtual Training Center, our recommendation is to keep the sub-function categorization as "Employee Performance Management." The NNSA Los Alamos Site Office has informed the NNSA Program Office that it does not fund these two investments through Human Capital Management. Therefore, the Program Office feels the current sub-function is correct.

We also believe the example of the EM Headquarters Central Internet Database (CID) being miscategorized on page 20 should be deleted from the report. As representatives from EM explained in a meeting with the GAO when this investment was raised, the CID exists as part of a DOE/Department of Justice settlement agreement with the Natural Resources Defense Council. The system is operational, as required by law, indefinitely or for ten years following the conduct of a second public meeting. EM maintains the database for public access, spending approximately $20,000 annually. It does not duplicate any functionality that exists in other IT investments and is a small business expense in relation to the other multi-million dollar investments cited in the report. While we agreed the CID could be re-categorized, we do not

Appendix V: Comments from the Department
of Energy

necessarily believe it should be. However, we changed the codes for the EM Headquarters CID and Office of Nuclear Energy Idaho National Laboratory Classified Cyber Life Cycle Management investments when we resubmitted the BY 2013 Passback Exhibit 53 Submission to OMB on January 10, 2012. Therefore, we believe this recommendation to be closed.

Again, thank you for the opportunity to review this report. If you have any questions related to this letter, please feel free to contact me at (202) 586-0166.

Sincerely,

Michael W. Locatis, III
Chief Information Officer

Appendix VI: Comments from the Department of Homeland Security

Homeland
Security

January 13, 2012

Mr. David A. Powner
Director, Information Technology Management Issues
U.S. Government Accountability Office
441 G Street, NW
Washington, DC 20548

Re: Draft Public Report GAO-12-241, "INFORMATION TECHNOLOGY: Departments of
Defense and Energy Need to Address Potentially Duplicative Investments"

Dear Mr. Powner:

Thank you for the opportunity to review and comment on this draft report. The U.S
Department of Homeland Security (DHS) appreciates the U.S. Government Accountability
Office's (GAO's) work in planning and conducting its review and issuing this report.

The Department is pleased to note that GAO did not identify any potentially duplicative
information technology (IT) investments at DHS in its sample. We appreciate GAO's
positive acknowledgement of the actions DHS has taken to identify and eliminate duplicative
investments and the tangible results which are expected to yield more than $40 million in cost
savings. The Department remains committed to eliminating any duplication in its IT
investments as part of its efforts to ensure taxpayer dollars are spent effectively and
efficiently.

The draft report contained one recommendation directed to DHS, with which the Department
concurs and has already implemented. Specifically, GAO recommended that the Secretary of
Homeland Security direct the Chief Information Officer (CIO) to:

Recommendation: Correct the miscategorizations for the DHS investments we identified
and ensure that investments are correctly categorized in agency submissions.

Response: Concur. The DHS CIO has corrected the miscategorizations as suggested, and
will correctly categorize these investments in future agency submissions. Corroboration of
these changes was previously provided to GAO under separate cover. DHS considers this
recommendation as implemented and closed.

Again, thank you for the opportunity to review and comment on this draft report. Technical comments were previously provided under separate cover. We look forward to working with you on future Homeland Security issues.

Sincerely,

Jim H. Crumpacker
Director
Departmental GAO-OIG Liaison Office

2

Appendix VII: GAO Contact and Staff Acknowledgments

GAO Contact	David A. Powner, (202) 512-9286, or pownerd@gao.gov
Staff Acknowledgments	In addition to the individual named above, the following staff made key contributions to this report: Shannin O'Neill, Assistant Director; Cortland Bradford; Javier Irizarry; Lee McCracken; and Kevin Walsh.